YOUR KNOWLEDGE HAS VALUE

- We will publish your bachelor's and
  master's thesis, essays and papers

- Your own eBook and book -
  sold worldwide in all relevant shops

- Earn money with each sale

Upload your text at www.GRIN.com
and publish for free

Simon Valentin

Aus der Reihe: e-fellows.net stipendiaten-wissen

e-fellows.net (Hrsg.)

Band 1521

Perception in Visual Communication. Evolution and Neurology of Vision

GRIN Publishing

Bibliographic information published by the German National Library:

The German National Library lists this publication in the National Bibliography;
detailed bibliographic data are available on the Internet at http://dnb.dnb.de .

Imprint:

Copyright © 2015 GRIN Verlag, Open Publishing GmbH
Print and binding: Books on Demand GmbH, Norderstedt Germany
ISBN: 978-3-656-95604-4

This book at GRIN:

http://www.grin.com/en/e-book/299245/perception-in-visual-communication-evo-
lution-and-neurology-of-vision

Karlsruher Institut für Technologie
ZAK | Zentrum für Angewandte
Kulturwissenschaft und Studium Generale

Visual Communication: Perception
Evolution and neurology of vision

Written assignment for the seminar:
Visual Communication and Culture

Winter semester 2014/2015

„The fact that people are born with two eyes,
but only with one mouth, suggests that they should see twice
as much as they should talk.“

Marie Marquise de Sévigné
(1626 – 1696), French aristocrat and writer

Contents

1. Introduction

Today, in our fast moving, computer-driven lives we are exposed to a myriad of sensations every second and not only from the environment or nature around us but also from ceaseless attacks through our modern media, that is mostly based on visual stimulation. So the topic of perception is, although it has been dealt with throughout the centuries from the stoic philosophy of the Greeks to modern neurologists, a highly current one that affects us all and every day in an unprecedented way.

Not least as a popular American crime drama television series with the title "Perception", where an eccentric neuropsychiatrist uses his unique perception abilities to solve complex criminal cases and a modern theatre play "Molly Sweeney" by Brian Friel, on stage at the moment at Theater Lindenhof in Melchingen, Germany, where the protagonist, a young woman, regained her eyesight through an operation and could not cope with the overwhelming sensations, show the current fascination of the topic.

In the following chapters I will draw attention to the basic principles of perception, especially visual perception as well as the evolution, concept and the functioning of our eyes to come to a better understanding of how we see things and the way our visual perception works.

2. Basics of perception and vision

2.1 What is perception?

The Cambridge Dictionary gives in fact two different meanings for this term: Firstly, perception is described as "the quality of being aware of things through the physical senses, especially sight" (Cambridge Dictionaries Online, n.d.) and it adds that drugs can alter our perception of reality. Secondly, the term perception also means belief, a belief or opinion that many people hold based on how things seem to be. For example can photographs or a film affect people´s perception of war or other events in the past (cf. Cambridge Dictionaries Online, n.d.).

These two definitions do not exclude each other. We rather experience perception as a process starting with a stimulus and leading to a person´s subjective understanding of the world. But this is not a simple process; it involves the processes of identification, interpretation and organization of information (cf. Goldstein, 2002, p. 1-3). In other words, perception is our sensory experience of our environment and it involves the recognition of environmental stimuli as well as the action taken in response to these stimuli. The perceptual process also delivers information about elements of the environment that may be critical to our survival. Thus perception "creates our experience of the world around us, it allows us to act within our environment" (Cherry, 2014).

Perception is experienced predominantly by the five classical senses of human beings: Sight, Touch, Smell, Taste and Sound. These are completed by a multitude of other senses like the feeling of hot and cold, that means temperature, the capability of keeping balance, for example when standing on one leg or by riding a bike, the feeling of pain and the kinaesthetic sense, the awareness of muscular movement and position, that tells you for example which hand is closer to the telephone when it rings or helps to find your mouth when eating with knife and fork or let you touch your nose with your finger even with closed eyes. Moreover, there are many internal senses which are for example responsible for breathing or the gag reflex or the sense of time (cf. Campenhausen, 1993, p. 4).

As the definitions above imply, perception is not just a passive receipt of signals, but it is the result of an active work of our brain, and as we will see influenced by memory such as recognizing a familiar person in the street, expectation and attention.

But before we immerse into the details of the perceptual process, which will be explained in the fourth chapter, I want to draw attention to some fascinating facts about vision in general and give a short overview over the process of evolution of vision and over the structure of the human eye and how it works.

2.2   Facts about vision

"Eyes are more exact witnesses than ears," (Diels/Kranz, 1992, B101a), stated the Greek philosopher Heraclitus and interprets vision as more important than all the other senses. There are different types of humans, who recipe their environment in different ways, some are more auditory, some more kinaesthetic or auditory-digital or even olfactory orientated. But for all of them vision is very important and for the big majority of human beings seeing is the predominant way to get information about the surrounding (cf. Thompson-Schill et al., 2009; Walter, 2013, p. 25-29).

This is not surprising as the human eye and the associated neurological system is one of the most complex organs and is able to distinguish about 50,000 shades of grey and 10 million colours and can in ideal conditions see up to 576 megapixels For comparison, an average camera is equipped with only 12 to 20 megapixels (cf. Scriba, 2007, p. 1-2). To evaluate this data, each eye sends about one gigabyte of information to the brain per second and our brain is able to process it in just a fraction of a second which requires about half of the brain to get involved (cf. Gröne, n.d.).

3. The evolution of vision

Picture 1: Owl with camera eyes

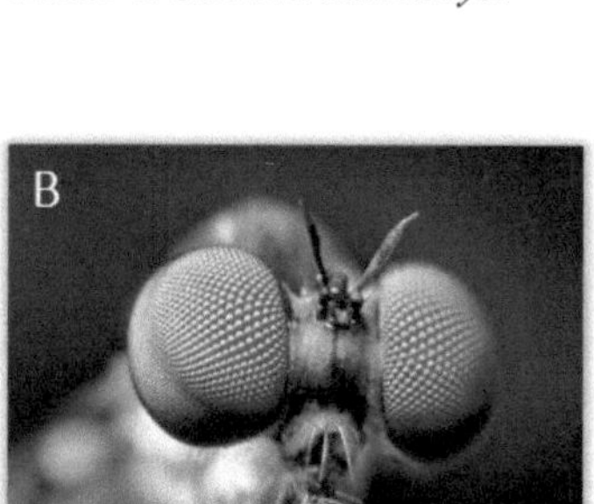

Picture 2: Robber fly with facet or compound eyes

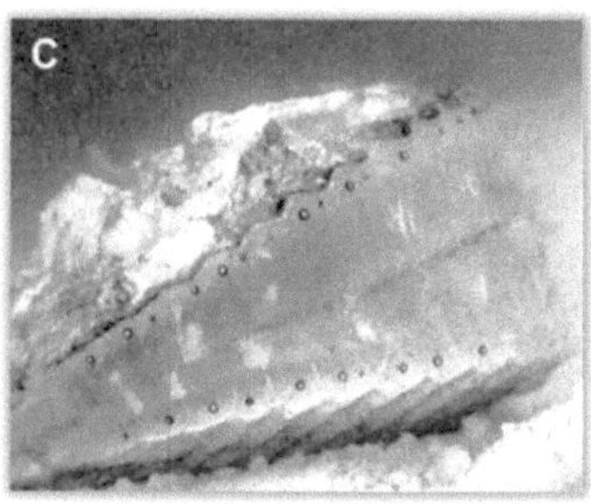

Picture 3: Scallop with reflector eyes

Animals as well as humans use eyes for their visual perception. Looking at animals we can discover a wide variety of different kinds of eyes (about 40 to 60 different kinds). Pictures 1 to 3 show the three most popular forms: The camera eye of the vertebrates with a lens (1), the complex facet eye of insects (2) and the mirror eye of the scallop (3) (cf. Goldstein, 2002, p. 216; Gehring, 2012).

In his book "On the Origin of Species", published in 1859, in which Charles Darwin developed his theory of evolution he wrote a whole chapter about the eyes and the ability to see. In the years of research he discovered and was fascinated about the different types of eyes of animals.

He had a problem explaining the existence of the sense of sight, because for him it appeared unlikely that something as elaborated as for example the eye of an eagle has developed just because of genetic variation and natural selection. With regard to the different types of eyes working completely differently he found it even more unlikely that every single sort of animal with special eyes has accidentally brought forth a unique kind of eye (cf. Gehring, 2012).

For a long time and sometimes still today the existence of the eye was used as a proof for an intellectual design and thus the existence of god instead of evolution and natural selection.

Darwin could not fully explain the development of the eye. He himself felt that something was missing. But he was convinced by his theory of evolution so at some point the eye must have developed. Because he could not believe that the incident of creating an eye has happened independently for more than once, Darwin´s thesis was that at a very early point of the evolution there was a very simple form of an eye consisting of just two cells that provided a selective advantage and from this prototype all different eyes have developed. He could not prove this thesis, but actually he was right (cf. Gehring, 2012; Goldstein, 2002, p. 127).

Many scientists after him took another view of it. Until 1995 the common opinion was that the eye in fact developed in a convergent way so at least 40 to 60 times independently from each other. But in 1995 the Swiss developmental biologist Walter Gehring discovered the master gene Pax6 which seems to be the starting point for the development of the eyes of all animals. With a multitude of experiments Gehring and his team could show that with this master gene they could control the eye development in all conceivable kinds of animals. So, with different mutations of the gene, they could grow mice without eyes when they deactivated the gene and even flies with additional and functioning eyes on different parts of their bodies, for example on their legs or on their wings.

The results proof that the programme for developing eyes is the same within all animals and human beings and all different kinds of eyes have indeed evolved from one point. Thus, Darwin´s thesis was perfectly confirmed, if only 100 years after his death (cf. Gehring, 2012).

4. The concept of vision

4.1 Structure and function of the human eye

As we have seen in chapter three there are different types of eyes, but in all cases vision is based on light and optical refraction. We will now take a closer look at how the process of seeing actually works and for that we will concentrate on the human eye. Picture 2 shows the most important parts of the anatomy of the human eye.

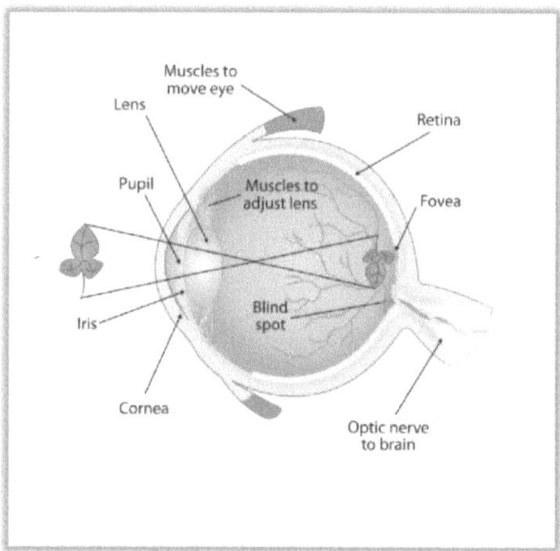

Picture 4: The human eye

Firstly, there are muscles around the eye that control the angle and are the most active muscles in the human body. The transparent cornea protects the pupil and the iris and helps together with the lens to focus the light, we will see that later. Directly behind the cornea is a coloured ring-shaped membrane called the iris. The iris with its adjustable circular opening, the pupil, regulates the amount of incoming light like a shutter. After passing through the lens the light must travel through the vitreous humour before it meets the retina. The retina consists of millions of light sensitive cells which transform the light energy into an electrical nerve signal which then travels to the brain via the optic nerve and eventually forms an image (cf. Stangor, 2012).

There are two classes of cells, rods and cones. Rods are good for black and white vision, they help us to distinguish between light and dark, cones are there for more detailed and colour vision. They need more light to get activated. This is why humans hardly see colours and have only grey scale vision in twilight conditions, when mostly the rods are working. Cones are situated in an area of the retina called fovea. The spot where the optic nerve exits to the brain is called blind spot, because there are no receptors on the retina. When light falls on this spot a potential image cannot be seen (cf. Campenhausen, 1993, p. 137; Goldstein, 2002, p. 39-45).

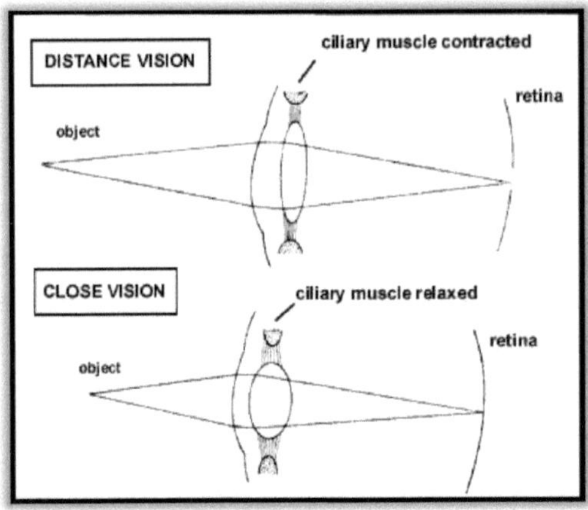

Picture 5: The path of light through the eye

As we can see in Picture 5 the elastic lens makes it possible to switch between near and far vision by adjusting the refraction of the light. If the focusing does not work properly, sharp vision is not possible and the result is either short- or farsightedness (myopia or hyperopia). About 60% of the population in the western countries need some form of vision correction. These people need glasses or contact lenses as additional lenses so the focal point hits the retina (cf. Florida Lions Foundation for the Blind, 2013; Stangor, 2012). Today an artificial lens or laser treatment can help to regain normal eyesight by changing the index of refraction of the cornea, so that the total refraction is correct again. Studies from Asia, the USA and Europe show that the percentage of short-sighted people in industrialized nations increases dramatically. In India, for example, the proportion is only 16%. Longer education and training periods and the increasing use of computers also in leisure cause the fact that the eyes simply lose the ability to see at a long distance. But at the same time development brings ever better ways to balance a defective vision (cf. Kuratorium Gutes Sehen, 2011).

The fact that we as humans like all vertebrates have two eyes, which leads to binocular vision, enables us with the help of the brain to see objects in a three-dimensional-manner and also judge distances which was a big advantage and helped us to survive in the evolutionary process.

4.2   Neurology of vision

Vision is one of the most complex activities that our brain conducts represented in every major area of the brain from the midbrain to the brainstem. As we have seen in chapter 2.2 a huge amount of information is processed, but how does the neural process work?

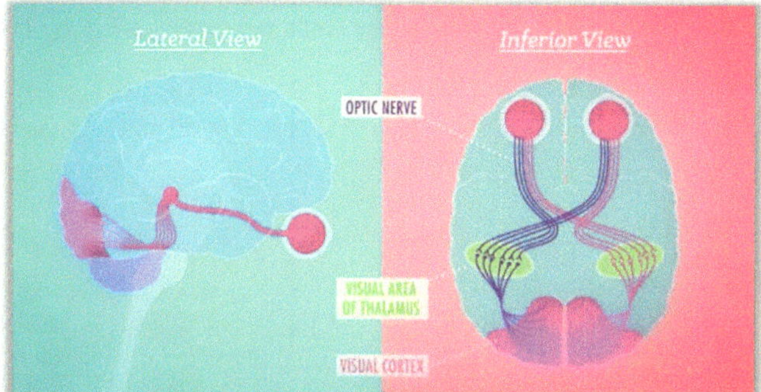

Picture 6: Way of signals from eyes to brain

In picture 6 we get an insight into the way the brain processes the optic signals. An incoming optic signal travels from the eyes through the thalamus to the visual cortex in the back of the brain. As we have seen in picture 2 the image on the retina is upside down from the actual image in the environment and the brain has to put it in the right way. The picture also shows that the signals are carried from the left eye to the right side of the brain and vice versa (cf. Green/Green, 2014; Stangor, 2012).

According to today's thinking there are different areas for recognizing different aspects of things within the cortex. If one region is not working properly, after a stroke for example, the person cannot "see" these things. It is also regarded an illness called prosopagnosia when people cannot recognize faces. They can see them with their eyes but their brains cannot process the information because one region in their brain is not working properly.

Now we want to go more detailed in the psychological perspective of visual perception. To better understand the way the perceptual process works it has been broken down into eight steps. The perceptual process is a psychological approach to make visual perception more understandable. It states that this process is working constantly and automatically that means we are not consciously aware of what happens in our brains and it is more than just a passive conversion of physical or chemical stimulation of the sense organs into signals for the nervous system (cf. Goldstein, 2002, p. 4).

1) The environmental stimulus

Our environment is full of stimuli that attract our attention. Imagine you ride your bike to work. On your way you pass a tram crammed with people, in the street a little boy is having an ice-cream or you pass a newly planted flowerbed in the park you cross every day. We perceive sound, light, temperature, smell et cetera. Normally the amount of stimuli is so overwhelming that it is not possible to notice everything but all these things serve as a starting point for the perceptual process (cf. Goldstein, 2002, p. 4; Cherry, 2014).

2) The attended stimulus

You are in the park and suddenly your eye falls on an object that you find particularly interesting and that attracts your attention, for example the newly planted roses in the park. The rose represents the attended stimulus and the visual progress starts from this point. The attended stimulus is the research object of all psychologists occupied in advertising. If they can find out what attracts people´s attention they are able to influence them (cf. Cherry, 2014). We know numerous examples from advertising strategies.

An interesting study was carried out in America with a group of American and Chinese students. The aim was to find out if the two culturally differently conditioned groups are attracted by different objects. And surprisingly they are. By tracking their eye movements, scientists found that Americans focused more on single objects in the foreground while Chinese focused more on a complex situation in the background. One example was a tiger in front of a forest scene. The explanation was that Asians live in complex social structures with a strict role situation. So for them it is more important to concentrate on the context and relations, while Americans or people in the western culture focus more on their independence and social networks are less important (cf. Chua, 2005, p. 1-5).

3) Reception or the image on the retina

The next step is that the attended stimulus delivers an image on the retina. As we could see in picture 4 and picture 5 the light passes through the cornea and pupil onto the lens of the eye thus projecting an inverted picture on the retina. But at this stage we are not yet aware of the perception but the receptors are stimulated (cf. Goldstein, 2002, p. 4).

4) Transduction

Transduction is the transformation of one form of energy into another form. In this step the image on the retina is transduced into electric signals. Now the visual message can be transmitted to the brain. The retina contains photoreceptive cells, the rods and cones, which contain a molecule called retinal. This molecule is responsible for transforming the light stimulus into visual signals that can be transmitted via nerve impulses (cf. Cherry, 2014).

5) Neural Processing

This is the process where the nerve signal travels through a highly complex system of pathways from the eyes to the visual area of the thalamus divided by sides and finally the electrical signal goes back to the visual cortex. We already discussed the transmitting process via neurons earlier in this chapter (cf. Goldstein, 2002, p. 5).

6) Perception

Only now we actually notice the stimulus object in the environment that is that we have become aware of the fact, to consider our previous example, that there is something in the flowerbed that has attracted our attention. At this point we have become consciously aware of the stimulus. But we are not yet fully aware of what it is and how to handle it (cf. Cherry, 2014).

7) Recognition

At this stage our brain interprets the stimulus and categorizes it. Recognition means the ability to interpret and give meaning to an object. It is at this stage that we recognize the rose as a rose in the flowerbed. The stage of recognition is essential for us because it gives us the chance to make sense of the world around us (cf. Cherry, 2014). "By placing objects in meaningful categories we are able to understand and react to the world around us." (Cherry, 2014)

In both the perception and the recognition information in our brain, like memories, conscience, knowledge or experiences play an underestimated role. Perception is not a neutral, objective process. Even the perception of things, the way how we see them is highly influenced by our brain. So our vision is so not so much a picture of the environment but rather what our brain does with the image of our environment (cf. Goldstein, 2002, p. 7-11).

8) Action

The last step of the perception process involves some kind of reaction to the environmental stimulus. This can be a motor action like running towards a recognized familiar person or like in our example to bend down and take a smell of the rose because we know from our memory that most roses have a pleasant fragrance (cf. Cherry, 2014).

The action is a result of the process of judgement. The step from recognising something to acting somehow is again influenced by information stored in our brain. Fundamental personal values are significant, for example fears, our risk tolerance or safety need. They determine how we judge and finally react to a special situation.

5. Conclusion

Visual perception is a complex process carried out unconsciously by our eyes and brain. But at several stages in this process it can be influenced. The process of perception and especially the process of vision is not a passive and objective absorption and representation of stimuli from our surrounding as we usually think, but is a highly subjective process. Our picture of our environment and thus our reality is to a large extent a product of our brain.

Especially advertising uses psychological tricks to attract our attention through colours, moving pictures and beautiful and interesting people. Through this process we are often encouraged to consume products we do not really need or we behave and even think in an intended way as we are prone to hidden persuaders.

Another point is our cultural background, our education and experience. Through pictures for example or news or all kinds of stimuli through the mass media with reminiscences to images in our brain we can easily be misled and sometimes we even fool ourselves.

So in conclusion I think we all should be more aware that we are in fact exposed to influences every day. Nevertheless we should stay open-minded for new pictures, new ideas and be able to sometimes leave old categories to embrace new ones in order to create and develop our own personality. In any case we should keep our fantastic brain occupied and let it develop every day.

Bibliography

CAMBRIDGE DICTIONARY ONLINE (n.d.): English definition of "perception", [online] http://dictionary.cambridge.org/dictionary/british/perception [05.02.2015].

CAMPENHAUSEN, Christoph von (1993): Die Sinne des Menschen - Einführung in die Psychophysik der Wahrnehmung, Stuttgart.

CHERRY, Kendra (2014): Perception and the Perceptual Process, [online] http://psychology.about.com/od/sensationandperception/ss/perceptproc.htm [02.02.2015].

CHUA, Hannah Faye / BOLAND, Julie E. / NISBETT, Richard E. (2005): Cultural variation in eye movements during scene perception, in: PNAS (Bd. 102, S.12629).

DIELS, Hermann / KRANZ, Walther (1992): Die Fragmente der Vorsokratiker, Weidmann Verlag, Berlin.

FLORIDA FOUNDATION FOR THE BLIND (2013): The Human Eye, [online] http://www.floridalionsfoundation.org/Eye.html [02.04.2015].

GEHRING, Walter Jakob (2012): Auge um Auge - Entwicklung und Evolution des Auges, [online] http://www.science-blog.at/2012/10/auge-um-auge-entwicklung-und-evolution-des-auges [04.02.2015].

GOLDSTEIN, Eugen Bruce (2002): Sensation and Perception, 6th. ed., Wadsworth-Thomson, Pacific Grove.

GREEN, Hank / GREEN, John (2014): Video-Crash-Course Sensation & Perception - Crash Course Psychology, [online] https://www.youtube.com/watch?v=unWnZvXJH2o, [27.01.2015].

GRÖNE, Torsten (n.d.): Unverstandenes Genie: Das visuelle Gehirn, [online] http://www.tg8.eu/webseiten/gehirn1.html [04.02.2015].

KURATORIUM GUTES SEHEN (2011): Kurzsichtigkeit auf dem Vormarsch – Korrektionsmöglichkeiten auch, [online] http://www.sehen.de/newsroom/kurzsichtigkeit-auf-dem-vormarsch-%E2%80%93-korrektionsmoglichkeiten-auch [05.02.2015].

SCRIBA, Jürgen (2007): Perfektes Auge - Blick in die Zukunft, in: ZEIT Wissen 04/2007.

STANGOR, Charles (2012): Introduction to Psychology – Seeing, [online] http://www.peoi.org/Courses/Coursesen/psy3/ch/ch4b.html [05.02.2015].

THOMSON-SCHILL, Sharon (2009): Visual Learners Convert Words to Pictures in the Brain And Vice Versa, University of Pennsylvania News, [online] http://www.upenn.edu/pennnews/news/visual-learners-convert-words-pictures-brain-and-vice-versa-says-penn-psychology-study [04.02.2015].

WALTER, Peter (2013): Über die Bedeutung des Sehens - Bemerkungen zur Physiologie und Psychologie des Sehens und Nicht-Sehens, Vorlesung RWTH Aachen, Universitätsklinikum, [online]http://www.ukaachen.de/fileadmin/files/klinik-augenheilkunde/bedeutung_von_sehen.pdf [06.02.2015].

List of illustrations